Björn Linnemann

Economy and transport in the Nile Region

Is the Nile a shaping element?

GRIN Verlag

Bibliografische Information der Deutschen Nationalbibliothek:

Die Deutsche Bibliothek verzeichnet diese Publikation in der Deutschen Nationalbibliografie; detaillierte bibliografische Daten sind im Internet über http://dnb.d-nb.de/ abrufbar.

Imprint:

Druck und Bindung: Books on Demand GmbH, Norderstedt Germany
ISBN: 978-3-656-25028-9

This book at GRIN:

http://www.grin.com/en/e-book/198593/economy-and-transport-in-the-nile-region

Economy and transport in the Nile Region

–

Is the Nile a shaping element?

Seminar:

Probleme, Konzepte und Methoden der Mensch-Umwelt-Interaktion

Universität Hamburg
Institut für Geographie

Author:

Björn Linnemann
M Sc Geographie

Submitted: 30.07.2010

CONTENT

LIST OF FIGURES

1 Aim and methodology

The aim of this paper is to analyze the economies and transport infrastructure of the countries in the Nile Basin. I will take into account latest economic data from the World Bank in conjunction with reflecting relevant literature. I will to take a closer look at the economies and the transport infrastructure of the Nile riparians. This way I want to point out the differences and certain characterisitcs of the Nile riparian states in regards to their economic condition and infrastructural situation. In order to integrate this topic into the wider scope of human nature interaction I will take into account the interplay of the economic factors with other aspects. I will take a broad perspective considering aspects of politics and social development as well as historic developments and future prospects. In doing so I want to figure out, which economic and infrastructural role the Nile waterways play and how the river actually interlinks the national economies, making them dependant on each other.

2 Introduction

Since ancient times waterways were of extraordinary importance for human civilization. Early civilization settled and developed close to waterways and in adjacent valleys that provided water and fertile soil for agricultural development. The availability of water is directly linked to the fate of a civilization. Rivers provide water for drinking, for food production, for energy and for transport. Consequently waterways have always been a central element of the economic environment and a factor of strength (Grey and Sadoff, 2002). The Nile River has eversince been of strategic and economic interest for many actors, for example in times of colonial control.

The Nile River and its water shed is shared by ten countries, namely Burundi and Rwanda, where the river has its source and Tanzania, Kenya, Uganda, the Democaratic Republic of Congo, Eritrea, Ethiopia, Sudan and Egypt, where the Nile flows to the Mediterranean Sea (see Figure 1). All these countries differ in their political and economic situation and they all do play a different role in the Nile region. They act from different political backgrounds and heritages and have different prerequisites in terms of natural resources. Nevertheless they do share the Nile as a major resource for water.

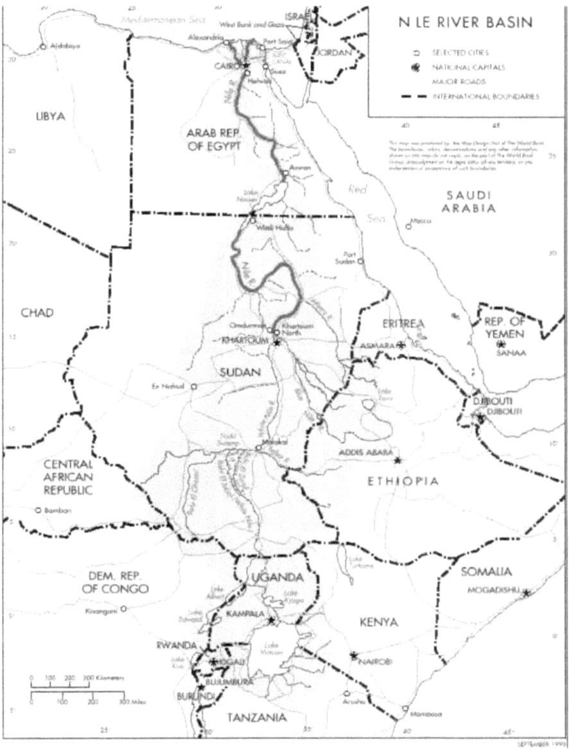

Figure 1: The Nile Basin
Source: www.africanwater.org/nile.htm

Today the Nile Basin is home to about one third of whole of Africa's population (Varis, 2000). This can broadly be explained by the significant importance of the Nile River for human development and for the capacity of the waterways to support the lives of so many people. The Nile basin is among the most critical regions of the world in terms of water resources (Varis, 2000). This makes the region extremely vulnerable to changes of the natural and environmental condition of the waterways. With the first effects of climate change in being noticed, the states in the Nile region face increasing challenges in terms of drought, environmental degradation, food security and socioeconomic development.

3 Characteristics of the economy in the Nile Basin

3.1 Overall economic characterisics of the Nile region

The economy in the Nile Basin is all in all quiet weak. The combined GDP of all Nile riparians was about 336 billion US$ in 2008. That is roughly the size of Denmarks economy, but with about 200 times the land area and 73 times its population (World Bank, 2010).

The latest World Bank data reaveals that the countries within the Nile Basin show highly variable economic performance and conditions and are very unequal in terms of economical capacity. Interestingly it looks like their overall economic capacity is not directly linked to the natural availability of water. Following the countries from downstream to upstream (from the Mediterrenan to the source of the river) one can see a clear decline in economic performance shown by the GDP data (see Figures 2 and 3). While the upstream riparians have the major share of the water supply and an abundant supply of rainfall, their economical condition is generally poor and they do not have either the economic, institutional and technical capacity to utilise their Nile waters appropriately (Allan, 1999). On the other hand the relative economically strong downstream states are mostly arid and depend completely on Nile water. So the countries with the most economic power are in general more waterscarce than the economically weak ones. Furthermore the economically weak states of the region have some of the highest poverty rates in the world (El-Battahani and El-Tom Hamad, 2005; Tesfaye, 2008). Four of the worldwide five countries afflicted by hunger in 2007 the most were located within the Nile Basin, namely Burundi, the Democratic Republic of Congo, Eritrea and Ethiopia (Welthungerhilfe, 2007). With the exception of Egypt and Kenya all of the Nile riparians are among the category of the 47 least developed countries world-wide (Mason, 2004).

The population of the Nile Basin countries has grown steadiliy and continues to do so. This means an increasing need of water for human consumption and economic activity in the future (Whittington and McClelland, 1992). Furthermore a plus of population means a growing need for food and thus more agricultural efforts to feed these people.

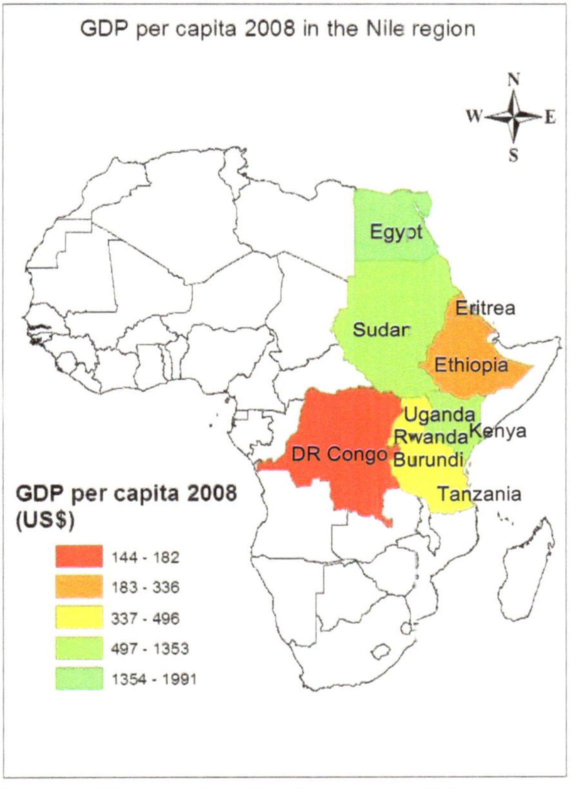

Figure 2: GDP per capita in the Nile region in 2008
Soure: own map based on World Bank, 2010

Water is a key resource for economy as well as for civilization itself. It is needed for nutrition, agriculture, livestock farming, energy generation, fishery and transport (Tesfaye, 2008). Taking a closer look at recent World Bank data (see Figure 3) reveals the general tendency that the Nile riparians share of agriculture in GDP grows with the natural availability of water. On the other hand it is easy to see, that the more upstream a country is, the less industry it has.

Having been under Sovjet influence many of the Nile region's countries share a socialist history, leaving them more or less left behind in the world economy with an economical burden. The complex ethnic composition in sub-Saharan Africa and borders mostly drawn by colonial rule further worsens the fundament for economic success.

	Egypt	Sudan	Ethiopia	Eritrea	Uganda	DR Congo	Kenya	Tanzania	Rwanda	Burundi
population , total (millions)	81,53	41,35	80,71	5,00	31,66	64,21	38,53	42,48	9,72	8,07
GDP per capita (current US$)	1990,53	1352,59	316,98	335,69	452,55	181,59	783,04	496,42	458,49	144,04
GDP (current US$) (billions)	162,82	58,44	26,49	1,65	14,53	11,59	34,51	20,49	4,46	1,16
GDP growth (annual %)	7,10	8,30	11,30	2,00	9,50	6,20	3,60	7,50	11,20	4,50
Agriculture, value added (% of GDP)	14,00	26,00	43,00	..	23,00	..	21,00	..	35,00	..
Industry, value added (% of GDP)	36,00	34,00	13,00	..	26,00	..	13,00	..	12,00	..
Services, etc., value added (% of GDP)	50,00	40,00	45,00	..	52,00	..	65,00	..	53,00	..
FDI net inflow (BoP current US$) (millions)	9494,60	2600,50	108,54	36,00	787,89	1000,00	729,04	744,02	103,35	3,83

Figure 3: Economic data of the Nile watershed region
Source: own table based on World Bank, 2010

The closer consideration of the national economies of the region will explain their dependency on the Nile waters. Nevertheless it should be mentionend that the available data has its limits. Firstly data is not easy to collect in unstable states and is often fractional. Secondly it is not capable of capturing the informal sector, such as informal trade or industry. This sector also relies on water resources from the Nile and does not occur in any of the statistics. One can assume that this informal sector is fairly significant in the less developed countries. So to a certain extend the economic importance of the Nile waters cannot be assessed accurately.

3.2 The special role of downtream Egypt and the Sudan

Within the Nile region Egypt and the Sudan play a special role. Nevertheless Egypt can be seen as an extraordinary economic and political power. Compared to the other countries it is the politicaly most stable one, which is generally reflected in the relative superior economic performance and level of development (Varis, 2000).
Egypt always received significantly more foreign direct investment (FDI) than any other country in the region (see Figure 4). This is largely due to its strategic location and importance, which attract investment from countries with political and economic interests and their oil resources, which are mainly exported to Europe via a pipeline from the Sinai pensinsula to the Mediterranean. Two major economic reforms in the 70s and in the 90s liberalized the market and opened it for foreign investments.

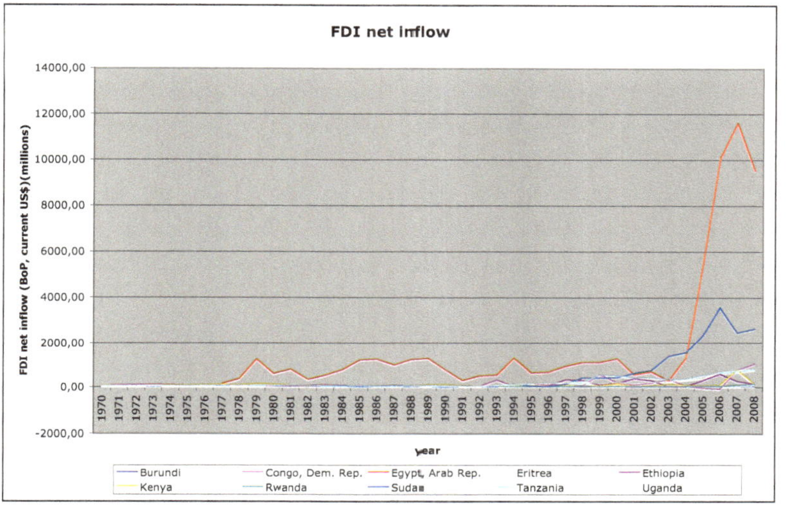

Figure 4: Development of FDI net inflows in the Nile region 1970-2008
Source: own figure based on World Bank, 2010

Due to the aridity in the Sahara and Lybian desert about 99 per cent of the population uses only about 5.5 per cent of the total land area and Egypt's habitable land is concentrated within the Nile Delta (Hamza, 1999; Milliman et al., 1989). Consequently most economic activity takes place in the highly fertile Nile valley. The economic structure and its major sectors depend on water resources provided by the Nile (Yates and Strzepek, 2000). The river provides more than 95 per cent of the total amount of water that is being used each year - with groundwater, agricultural drainage water and treated municipal water accounting for less than 5 per cent (Swain, 1997). This alone shows the extraordinary importance of the Nile waters for the country.

Compared to other Nile Basin countries Egypt has the lowest share of agriculture of the GDP. Even though the crop yield grew steadily, the share decreased over the last decades. The agriculture of Egypt can be divided into old and new lands. The old lands are along the Nile River and the delta, whereas the new lands are the newly reclaimed areas that used to be desert (Mason, 2004). Since 1960 the area used for agriculture has increased mainly because of application of hydrological technology such as the Aswan dam. The introduction of hydrological technology allowed permanent human settlement and economic activity far from water (Grey at al., 2002). Though such projects, irrigation schemes and connected desert reclamation projects supported by the waters of the dam at Lake Nasser helped to foster

agriculture, farming capacity is still limited. Domestic farming is far from being sufficient to feed Egypt's people. Due to the high rates of population growth and rise of incomes Egypt is forced to import nearly 50 per cent of of its food (Sehmi, 1996).
The industrial sector is relativley large and growing thanks to massive foreign investments since the 1980s. In around 90 industrial zones the country has mainly food processing, textile, chemical, cement and steel industry as well as some manufacutring in the automotive sector (Egypt State Information Service, 2010). The biggest industrial areas are located around Alexandria, Helwan and in greater Cairo, thus being in close reach to natural water resources.
It is worth to take a closer look at the service sector, which accounts for half of the GDP and is the biggest economic sector in the country. Tourism accounted for 20% of the service sector in 2004 and is the major subsector within services and one of he most important economic sectors in the country (Ibrahim and Ibrahim, 2006). Tourism is the main source of foreign earnings. Egypt is targeting 14 million tourists by the end of 2011. That is more than twice as much as in 2000, when more than 5 million tourists came into the country (Research and Markets, 2010). Nile cruises are a major tourist draw, which means that besides the need for the Nile to be a pure water resource, the river is an important natural assett attracting tourists and consequently an influx of foreign currency. This provides us with a new persepective towards the relevance of the Nile for Egypts economy. Thanks to the successful irrigation schemes Nile water appeared to be plentiful in Egypt in the past, but this is coming to an end as population growth, climate change and resource pollution is putting pressure on the water supply situation and thus influencing the economic capacity. Egypt is expected to have 90 to 101 million people in 2025 (Tesfaye, 2008; Whittington and McClelland 1992).
Still far behind Egypt the Sudan comes second in terms of economic performance in the region. In the recent decades the Sudan was a place of massive political disorder, conflict and war. This is the main reason, why the country is economically in such a bad shape compared to its potential. The conflicts and the extensive war in parts of the country were and still are of very high cost in terms of human capital, natural and financial resources. These resources are not available for economic activity and development, let alone the instable and economically unfriendly environment the conlficts caused. Moreover as a result of its policy the country was suffering from an economic embargo for most of the 1990s and the beginning of this century. The Sudan is still considered to be one of the least developed countries in the world (El-Battahani and El-Tom Hamad, 2005). While showing constant growth rates in the last years, mainly due to increases in oil production, high oil prices and

large inflows of foreign direct investment related to recent oil exploitation (China is the main investor), the country's GDP is far smaller then the one of Egypt. Nevertheless the Sudan shares Egypt's fate of a negative trade balance. Imports have continuously been greater than exports (El-Battahani and El-Tom Hamad, 2005). The service sector share of 40 per cent is the lowest among all riparian coutries with available data. While the GDP share of industry is pretty much on par with Egypt (mainly agricultural processing and oil refining), the Sudan is more active in agriculture relative to its GDP, binding about 70 per cent of the economically active population. This is partly due to the countries higher precipitation rate and less evaporation in the south (Mason, 2004). The contribution of water to GDP is not shown in any figure, but can generally be derived from the GDP contribution of agriculture, which in recent years consumed more than 90 per cent of the water. Only 5 per cent of the water is used for municipal, human and animal needs and only 1 per cent goes to industry. Bearing in mind that agriculture provides raw materials for over 85 per cent of the manufacturing industries the dependancy of the whole value chain on water provided by the Nile becomes obvious. As for 63 per cent of the basin area lies within the Sudan's borders, the country plays a key role in the region together with Egypt (El-Battahani and El-Tom Hamad, 2005). Due to their aridity both countries are most dependant on water from hydrological systems. Mostly covered by the Sahara, the northern half of the Sudan and Egypt are either desert or semi-desert with very rare precipitation (Tesfaye, 2008). To satisfy their water needs both countries rely on hydrological systems and artificial irrigation. They have a common border and each of the countries is directly impacted by any activity related to the Nile in one of the states. Taking into account this situation Egypt and the Sudan have bilaterally negotiated the allocation of Nile waters between each other in the past. According to the 1929 and 1959 Nile Waters Agreements the two countries are allocated the entire average annual flow between themselves. All other Nile riparians were left out of the process. The latter agreement allowed the Sudan to consume about 22 per cent of the Nile flow that passes through it. The rest of the water was allocated to Egypt. The problem is that such an amount of water would only be sufficient to irrigate 5 per cent of the Sudan's potentially arable land. Nevertheless the country is not even capable to consume this amount of water with its current technical means (El-Battahani and El-Tom Hamad, 2005). Irrigated agriculture has an enormous economical potential. But the main concern is not insufficient water, but the capability in the upkeep of technology and management for efficient water use (Mason, 2004). In many cases the political disorder further worsened the situation, when for example attacks by Sudanese People's Liberation Army (SPLA) caused the

abandonment of the construction of the Jonglei Canal in the south (Swain, 1997).
The system of existing and proposed dams and reservoirs along the Nile and its tributaries could help to regulate flow and optimize water management. Consequently the agro-industry could improve its efficiency to meet the continuous and growing demand for food, drink and industrial goods (El-Battahani and El-Tom Hamad, 2005). Since the completion of the Aswan dam in 1970 it plays a key role in Egypt's agricultural capacity and its independancy from seasonal floodings. Furthermore with the nature of dams being multi-functional the scheme has provided the additional benefits of hydroelectricity on top of creating a reservoir for irrigation purposes. Besides all mentioned benefits the Aswan Dam for example as well as other dams had negative effects on the environment. By holding back nutrients from the further way downstream fisheries are impacted significantly (Hamza, 2006). Effects can even be seen in the Mediterranean, where the numbers of certain species of fish decline. This again has a negative effect on the fishery industry.
Though energy in Sudan is mainly derived from biomass and Egypt only generates a very small share of its electricity from hydrogical powerstations, hydroelectricity has a very high potential. There are several power stations along the Nile River. In the Sudan the Roseires Dam on the Blue Nile grid did account for the largest part of the country's hydropower before 2009. When the Merowe Dam in the north of the country was inaugerated that year, it doubled the country's total electricity production (El-Battahani and El-Tom Hamad, 2005). Even though they still mainly rely on oil and gas and biomass, the hydrological system in both Egypt and the Sudan is thus not only important for irrigation but as well for energy generation. A project worth to be mentioned in this respect is the Desertec power generation project, which aims to generate energy for the EUMENA regioen (Europe, Middle East and Africa) by mainly setting up and connecting a solar power network in northern Africa. It would most certainly make efficient use of natural energy resources, especially in terms of solar power. A dedicated use of hydropower though is rather little and concentrates on the Aswan dam. Nevertheless it has to be argued, whether the economic benefit of such a project will reach the African economies. The whole project, the funds and the technology and the end users are mostly European. Whether and to which degree a technology and knowledge transfer would to happen is left unanswered until now (Scholvin, 2009). If there would be benefits for the Nile region at all, these benefits would rather concentrate on the downstram countries and here especially Egypt.

3.3 The uptream countries – rich waters and poor economies

The region of the upstream Blue and White Nile is home to some of the poorest countries on earth. As has been stated before, the economic capacity tends to decrease upstream and most upstream Nile riparians indeed show comparably weak economical performance. In contrast to Egypt and the Sudan nearly all other countries are generally not water scarce, getting a sufficient share of rainfall thanks to their mostly tropical climate.

Many of the upstream countries share the heritage of extensive social unrest, armed confilcts and civil war. According to the World Bank (2000), one in five people in sub-Saharan Africa lived in a country affected by conflict. In ongoing conflicts starting in the 1960s in Burundi for example about 500.000 people lost their lives. Genocide in the 1990s in Rwanda killed about 10 per cent of the population. These conflicts not only were of high economic cost for the effected countries and therefore had and still have a negative effect on the economies (Lopez and Wodon, 2005). Some of these conflicts were at least partly caused by high inequalities of economic wealth among the citizens. Depending on the ruling government, economic decisions were often in favour of certain ethnic groups, which boosted unequalitiy and conflict potential (Ngaruko and Nkurunziza, 2000). The interplay of economic crises and conflict left many of the Nile's riparians devastated. This needs to be considered, when one looks at the economic and social situations in those countries.

Together with Eritrea Rwanda and Uganda are by far the smallest economies in the Nile region. Even though empirical data is sometimes incomplete, it can be said that in all of these economies services is the most important sector in regards to the share of the GDP, followed by agriculture (see Figure 3). Industry mainly concentrates on processing agricultural goods.

Agriculture usually employs the major share of people and is mainly small scale subsistence farming and for export. This economical structure makes the economies highly dependant on agriculuture. Being in close geographical proximity especially for Burundi, Rwanda and Uganda the export of aricultural goods such as coffee, tea or tobacco is the main source of foreign currency. Uganda, which does export some oil and gas as well somehow takes a special place as a recently comparably stable state and prospering economy (Kabanda and Kahangire, 1994). Nevertheless, being this dependant on exports is creating an overall vulnerability of the countries' economies to price shocks for certain goods. The economic reform in the Rwanda of the 1980s proposed by the World Bank for instance pushed the country's agriculture into crisis. When coffee prices plummeted, the farmers and the whole sufferd from a dramatic loss of income (Chossudovsky, 1996). There would be sufficient water for sustainable farming and other economical purposes in the region, but the political preconditions together with missing funds and technical know-how do not allow efficient use of the water resources.

Ethiopia is the second largest nation in Africa in terms of population and its economic situation is deeply interwoven with Eritrea. Both Ethiopia and the very small Eritrea have very weak economies compared to their size and population. The comparably small GDP rates can be explained with the aftermath of civil war in Ethiopia and the Ethiopian-Eritrean war from which both countries are still suffering. The two economies have a fairly low agricultural output and minimal exports. Economic growth cannot keep up with population growth and feed the people. Furthermore the population growth has degrading effects on the land further worsening the conditions for appropriate agriculture (Tesfaye, 2008). Neverthless both countries are in a prime geographical position with plenty of precipitation in the highlands. The largest amount of water from the cool Ethiopian and Eritrean highlands drains into the Blue Nile, making the highlands the main water feeder for the Nile region (Ayalew, 1999). Their proximity to rich sources of water provides good basic conditions for agriculture. The potential for sufficient agriculture is given, but due to poor irrigation practice and little availability of know-how technical means for water management, agricultural production is limited (Tesfaye, 2008).

Kenya and Tanzania show a slightly better economical performance, which can be explained with their broad coastline with access to the Indian Ocean as an advantage in terms of trade. Most agriculture takes place around Lake Victoria, but production is limited due to an arid inland and insufficient irrigation. Kenya is mostly arid and prone to drought, which is reflected in the little share of agriculture of the GDP. The overall economic conditions with an industrially polluted ecosystem and political unstability along with occasional unrests are disadvantgeous (Tesfaye, 2008). Tanzania depends heavily on rainfed agriculture, employing the largest share of the national workforce. Main export goods are coffee, tobacco, cotton and the infamous nile perch (CIA, 2010). The country furthermore experienced significant growth rates in tourism over the last decades (Neumann, 1995).

Even though the Democratic Republic of Congo is only peripherical to the Nile Basin, the picture that has been drawn by most of the sub-Saharan economies does continue here. Relative to its size and population the economic performance is extremely weak, even though it is built upon massive and rich natural resources.

Among the already politically weak states, Democratic Republic of Congo is extremely unstable and prone to massive conflict.

The whole region is strategic in terms of energy supply. Oil and gas is found throughout the region in varying amounts. Like in Uganda, some sources are still untapped. In terms of hydropower generation all countries are behind their potential. Poor maintanance has lowered performance of many hydropower stations. Only about 11 per cent of the estimated potential has been realized until now. In Ethiopia, less than 2 per cent of the hydroelectric power potential has been tapped (Mason, 2004) Nevertheless many countries do already rely on hydroelectricity besides biomass.

3.4 The correlation between water and economy

Mason (2004) analyzed the water withdrawal rates for the Nile Bains countries (see Figure 5). Comparing the water withdrawal rate and the economic strength in terms of GDP one can notice a tendency of these factors to correlate. The differences in the point of time the data has been collected are neglectable, given that the economic rank order did not change significantly since the time, when the waterwithdrawal rates have been determined. Egypt has the strongest economy and the highest water withdrawal rate followed by the Sudan in both aspects.

Country	Year (b)	Total freshwater withdrawal, km^3/year (a)	withdrawal / availability in % (a & table 5.3)	Estimated year 2000/capita withdrawal, m^3 (b)	Domestic use, % (a)	Industrial use, % (a)	Agricultural use, % (a)
Burundi	1987	0.10	2.8	14	36	0	64
DR Congo	1990	0.36	0.04	7	61	16	23
Egypt	1993	55.10	95.5	809	6	8	86
Eritrea							
Ethiopia	1987	2.21	1.8	31	11	3	86
Kenya	1990	2.05	6.8	68	20	4	76
Rwanda	1993	0.77	12.2	100	5	2	94
Sudan	1995	17.80	20.1	597	4	1	94
Tanzania	1994	1.16	1.3	35	9	2	89
Uganda	1970	0.20	0.3	9	32	8	60

Figure 5: water equivalent as percentage of water availability
Source: Mason, 2004

At the end of the scale one finds Burundi and presumably Eritrea both having the weakest economies and the lowest water withdrawal rate (There is no data on water withdrawal available for Eritrea. Nevertheless due to their relative comparability a rate in proximity to the one of Burundi can be expected). The same tendency can be found in the ration of water withdrawal and water availability. Here Egypt is exploiting the highest percentage of the available water, whereas the poorer countries tend to make use of only a fractional amount of the available water resources. Still these correlations are not permanent throughout all countries, but the tendency remains. Furthermore the predominent part of the water is used for agriculture throughout all countries with the exception of the Democratic Republic of Congo, where most water is used for domestic purposes.

4 Transport infrastructure in the Nile Basin

4.1 General condition of transport infrastructure

The transport infrastructure is a backbone of the economy. For trade, supply, labour commuting as well as import and export it needs support by sufficient transport utilities on land, water and in the air. Access to rivers and the ocean with harbours and water trade routes, roads and railroads connecting places as well as airports and air transport capabilities are critical factors for economic development.

Like the economic capacity the level of infrastructure tends to decrease upstream (see Figure 6). Railway systems (no data on rail line length available for Burundi, Eritrea and Rwanda) in the whole region are mostly a colonial heritage and generally in poor condition in the upstream counries. Only a fractional part of the roads in the upstream countries are paved.

For most African countries, distance from their primary markets and the high transport intensity and cost of their products are major impediments for production and trade (Esfahani and Ramirez, 2003). Bureaucratic and procedure at ports and border post further as well as frequent corruption hinder efficient trade (Amjadi and Yeats, 1995). Air transport is generally weak in the region. Nevertheless the highest numbers of air departures take place in Kenya, Ethiopa and Egypt, which can generally be explained with their national airlines being fairly competitive and the capital airports serving as hubs for intercontinental flights.

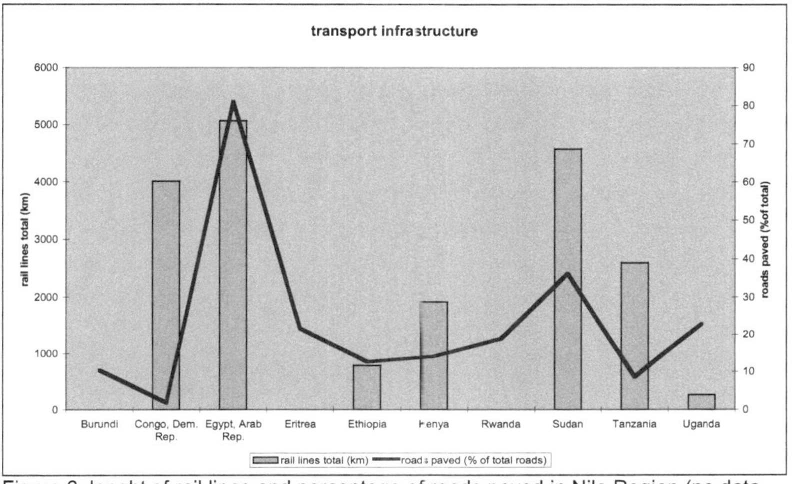

Figure 6: lenght of rail lines and percentage of roads paved in Nile Region (no data on rail lines for Burundi, Eritra and Rwanda)
Source: own graph, based on World Bank

The Nile is not shippable on a commercial scale in many sections, partly due to dams or other water schemes interrupting the flow or because of very shallow water. Nevetheless as for the Nile is a source of water and fertility, the following detailed analysis of the current transport infrastructure in the Nile region will clarify wheather and how strong it is linked to the Nile.

Like the economic data, particularly for the upstream countries the available data on infrastructure is sometimes incomplete and unreliable. The often very unstable states do not provide possibilities for an appropriate data collection. But even though the available data should be treated with caution, it does give hints about the infrastructural conditions.

4.2 Infrastructural advantages in downstream countries

As the leading economic and political power Egypt has the most developed transport infrastructure in the region. The country is in an extraordinary strategic location, having access to both the Mediterrenean and the Red Sea, which is the backbone of trade. It can be assumed that Egypt has the most port activity corresponding to its economic strength, even though data on transport infrastructure for all other countries is mostly unreliable or missing. The countries major airports are located

along the Nile or at the sea and the highest number of registered air carrier departurures each year makes the country the most important hub in the region.
The railroad infratructure, which beginnings date back to colonial rule, emphasizes the significance of the Nile as for the total 5063 kilometers of railroad mostly serve the densly populated Nile Delta all the way down the Nile to the Aswan dam (see Figure 7). The same applies for the road network. As of 2008 81 per cent of all roads were paved, mainly proceeding along the Nile from north to south and connecting outer reaches in the desert to the river. There is no evidence of the river being used for transport of goods nowadays, though being touristically important for Nile as has been mentioned before.

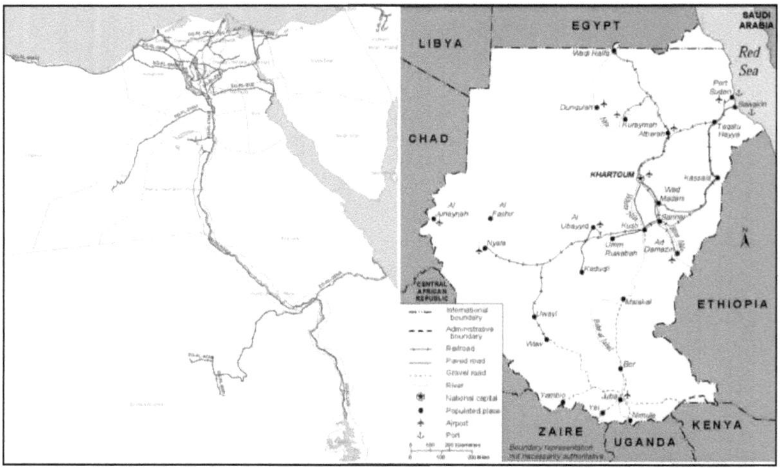

Figure 7: Railway system of Egypt and railway and road network of the Sudan
Source: http://www.euromedtransport.org/108.0.html and
http://lcweb2.loc.gov/frd/cs/sudan/sd03_08a.pdf

Similar to Egypt the railway system of the Sudan originated from military purpose in times of colonial rule and is generaly unefficient (Sellin, 1978). It is slightly shorter and runs from the Egyptian border to Khartoum following the flow of the Nile (see Figure 7). South of the capital it then reaches westwards, connecting the White and the Blue Nile and further into southern Darfur. The other branch reaches eastward to the Red Sea, connecting important Port Sudan to the center of the country and thus serving mainly for external trade (Sellin, 1978). The railway system does not reach further south leaving the largely autonomous south of Sudan without rail transport means.

The road network has the tendency to be denser in the north of the Sudan. Downstram and especially in lowlands and valleys floods of the Blue Nile frequently damage transport infrastructure like roads, which are moslty unpaved (36 per cent of all roads in 2001). This is the downsinde of the positive effects of floods on agriculture, renewing the land's fertility and recharge shallow aquifers (El-Battahani and El-Tom Hamad, 2005). Relative to its size air traffic in the Sudan is neglectable. Transport on the river never grew to significant size in the Sudan as well. Nevertheless the Nile was and still is, used to transport goods to different places along its long path on a very small scale and often for informal purpose. Some north-south roads serve the oil fields of central and south Sudan and there is a dedicated oil pipeline to Port of Sudan (Sellin, 1978).

4.3 Worsening transport infrastructure in upstream countries

Ethiopia and Eritrea both operate railway lines from their capitals to the Red Sea. Their roads are in a similar bad condition to the ones of the other upstream countries. Distance and limited coastal access all hinder a country's ability to exploit economies of scale, lowering its production efficiency (Sachs and Warner, 1997). Egypt, the Sudan, Eritrea, Kenya and Tanzania are the only countries among all Nile riparians that have sea access and sea harbours. Landlocked Burundi, Rwanda and Uganda share the fate that goods have to travel long distances compared to their regional partners Kenya and Tanzania. This high transport cost intensity of commodity trade as well as the level and scope of production and trade for the landlocked countries is limited (Limao and Venables, 1999; Obwona, 2001). The overall condition of transport infrastructure is another obstacle. Neither road nor railway transport systems are on a comprehensive and sufficient level (no data on rail line length available for Burundi, Eritrea and Rwanda available). Around 20 per cent or even less of all roads are paved and only Uganda has a short strip of railway connection from Kampala to the Kenyan border. From there the line continues to the Kenyan port of Mombasa at the Indian Ocean. Both Kenya and adjacent Tanzania have access to the Indian Ocean, but a poorly maintained road network and especially little paved roads in Tanzania. On the other hand Tanzania has 2600 kilometres of rail network connecting Lake Victoria and Tnaganjika Lake to the Indian Ocean. Rail and road networks are usually not well maintained in these countries due to missing funds and limited political commitment. Another important means of transport for

Kenya, Tanzania and Uganda are the Lake Victoria ferries, connecting the banks of the lake.

The Democratic Republic of Congo has a neglactable role in transport in the Nile Basin as for it is not connected to any other riparian by train and roads are usually unpaved and poorly maintained.

5 Conclusion and outlook

As we have seen, economy and infrastructure are only parts of the human natural interaction. Society and environement are interlinked. Economy and infrastructure as an outcome of societal activities are consequently interlinked to nature as well. The environment and the societal prerequisites are the very basis for economic and infrastructural activities as for they provide the ground upon which economy and infrastructure develops. At the same time economy and infrastructure are interlinked as well, because the quality of one factor influences the other factor (see Figure 8).

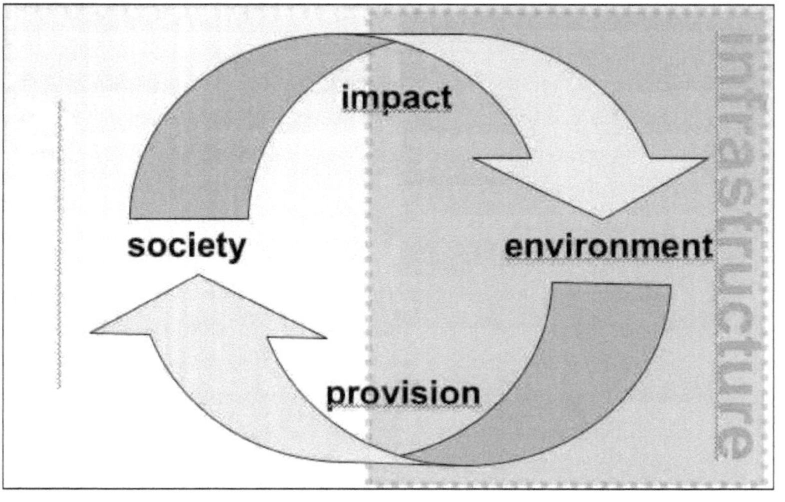

Figure 8: Human nature interaction system in economy and infrastructure
Source: own graphic

In the preceding chapters I have drawn correlations between the economic performance and the availability of water as well as between the economy and the dimension and the condition of transport infrastructure. Comparing all data a tendency towards a north-south-decline was discovered in economic as well as infrastructural terms. Furthermore we have shown that the Nile waters are a shaping element for the transport infrastructure. Taking into account the differences in economic performance and transport infrastructure it is self-evident that both the percentage of roads paved, road density and condition and reach of the railway and road network are positively correlated with economic productivity (Ndulu, 2006).

The Nile among other waterbodies acts as a lifeline, being the connecting element. Bearing in mind the upstream-downstream differences The Nile can be understood as a north-south axis. As for transport infrastructure and national economies depend on the Nile, the riparians find themselves being dependant on each other. As for they share the same resource, they are impacted by most of the action taken related to the Nile, no matter by whom.

Their economies are based on the availability of water. Nevertheless we have seen, that the natural availability of water per se does not form the basis for economic prosperity. Being largely water scarce the northern states like Egypt and the Sudan are in desperate need of the Nile's water provided from upstream. On the other hand the upstream countries do generally have rich water resources, but still limited water availability. There is a much greater potential to make use of the rich water resources, but fail they to do so, because they are lacking the know-how, the funds and the techniques. Better water management techniques would allow more water flow in the Nile. So one can say that the Nile region is not water scarce per se. The natural dispersion of water creates multilateral dependencies in this respect. Appropriate water management and technical means in the upstream countries would help to mitigate hydrologic risk (Grey et al., 2002). Not only would the weak upstream economies be able to use more water for a more efficient agriculture and development. The arid downstream countries would clearly benefit from this as well. It seems that only a common management of the water resources would help to cover everyones needs.

With better management, appropriate funds and maintanance the Nile Basin countries can generate power from hydropower stations and sell it. The potential to develop hydroelectric power in the Nile Basin is enormous. This could mean a massive economic benefit for water-rich, but poor countries like Ethiopia or Eritrea (Mason, 2004). Developing and linking the power systems could boost industrial activitywithin the Sub-basin, and power could be traded with neighboring countries, the Middle East and Europe (El-Battahani and El-Tom Hamad, 2005).

A common management of the Nile waters is on the way. All Nile Basin countries countries, with the exception of Eritrea are members of the Nile Basin Initiative (NBI), an organization founded in 1999 and aiming to jointly manage and develop the Nile waters. Generally the allocation of resources and their benefits can lead to confrontation and conflict. Taking into account the conflict-laden history and present of many Nile riparians a common management initiative is a major instrument to prevent conflicts and wars arising from the allocation of Nile water. The major problem and obstacle for the success of joint managmenet like initiatives like the NBI is and will be the high political instability of many countries. With the effects of climate change already felt especially inthe dry regions, immediate action is needed nonetheless.

REFERENCES

Ayalew, M. (1999): The Effect of Seasonal Rainfall on Landslides in the Highlands of Ethiopia, in: Bulletin of Engineering Geology and the Environment Volume 58:1, Berlin, pp. 9-19

Allan, A. J. (1999): The Nile Basin: Evo ving Approaches to the Nile Waters Management, in: Occasional Paper 20, SOAS Water Issues Group, London.

Amjadi, A. and Yeats, A. (1995): Have Transport Costs Contributed to the Relative Decline of Sub-Saharan African Exports?: Some Preliminary Empirical Evidence, Policy Research Working Paper Series 1559, World Bank, Washington.

Chossudovsky, M. ((1996): Genocide in Rwanda, in: Economic and Political Weekly, Vol. 31:15, Mumbai, pp. 938-941

El-Battahani, A. and El-Tom Hamad, O. (2005): Sudan and the Nile, in: Aquatic Science, 67, Dübendorf, pp. 28-41.

Esfahani, H.S. and Ramirez, M.T. (2003): Institutions, Infrastructure and Economic Growth, in: Journal of Development Economics, 70: 443-77.

Grey, D. and Sadoff, C. W. (2002): Beyond the River: The Benefits of Cooperation on International Rivers, in: Water Policy 4, The World Bank, Washington, pp. 389-403.

Grey, D., Sadoff, C. W. and Whittington, Dale (2002): Africas International Rivers: An Economic Persective, The World Bank, Washington.

Hamza, W. (1999) Economic Development and Compensatory Measures Related to the Management of the Egyptian Freshwater Resources. In: Water Security in the Third Millennium, Mediterranean Countries as a Case, Como, pp. 12-15.

Hamza, W. (2006): The Nile Estuary, in: The Handbook of Environmental Chemistry, Vol. 5, Part H, Berlin, pp.149-173.

Ibrahim, B. and Ibrahim, F. N. (2006): Ägypten, WBG-Länderkunden, Darmstadt.

Kabanda, B. and Kahangire, P. (1994): Irrigation and Hydropower Potential and Water Needs in Uganda – An Overview, in: The Nile: Sharing a scarce resource, Howell P.P. and Allan J.A. (eds), Cambridge, pp. 217-226.

Limao, N. and Venables, A.J. (1999): Infrastructure, Geographical Disadvantage, and Transport Costs, Policy Research Working Paper Series 2257, World Bank, Washington.

Lopez, H. and Wodon, Q. (2005): The Economic Impact of Armed Conflict in Rwanda, in: Journal of African Economies, 14, Oxford, pp. 586-602.

Mason, S. A. (2004): From Conflict to Cooperation in the Nile Basin, Zurich.

Milliman, J. D., Broadus, J. M. and Gable, F. (1989): Environmental and Economic Implications of Rising Sea Level and Subsiding Deltas: The Nile and Bengal Examples, in: Ambio, Vol. 18, No. 6, Lawrence, pp. 340-345.

Ndulu, B. J. (2006): Infrastructure, Regional Integration and Growth in Sub-Saharan Africa: Dealing with the Disadvantages of Geography and Sovereign Fragmentation, in: Journal of African Economies, 15, Oxford, pp. 212-244.

Neumann, R. P. (1995): Local Challenges to Glocal Agendas: Conservation, Economic Liberalization and the Pastoralists Rights Movement in Tanzania, in: Antipode 274, Sussex, pp. 363-382.

Ngaruko, F. and Nkurunziza, J. D. (2000): An Economic Interpretation of Conlifct in Burundi, Journal of African Economies, 9:3, Oxford, pp.370-409

Obwona, M. B. (2001): Determinants of FDI and Their Impact on Economic Growth in Uganda, The Economic Policy Research Center, Kampala.

Sachs, D. and Warner, A.M. (1997): Sources of Slow Growth in African Economies, in: Journal of African Economics, 6:3, Oxford, pp. 335-376.

Scholvin, S. (2009): Desertec: Wirtschaftliche Dynamik und politische Stabilität durch Solarkraft?, in: GIGA Focus Nr. 11, Hamburg.

Sehmi, N. (1996): The Enigmatic Nile, in: World Meteorological Organization Bulletin, 45:3, Geneva, pp. 267.

Sellin, H. (1978): Zur Politischen Ökonomie des Transportsektors in der Republik Sudan, in: Africa Spectrum, 13, Hamburg, pp. 35-44.

Swain, A. (1997): Ethiopia, the Sudan and Egypt: The Nile River Dispute, in: The Journal of Modern African Studies, 35:4, Cambridge, pp. 675-694.

Tesfaye, A. (2008): The Political Economy of the Nile Basin Regime in the Twentieth Century, New York.

Varis, O. (2000): The Nile Basin in a Global Perspective: Natural, Human and Socioeconomic Resource Nexus, in: Water International, Volume 25:4, London, pp. 624-673.

Yates, D. N. and Strzepek, K. M. (2000): Responses and Thresholds of the Egyptian Economy to Climate Change Impacts on the Water Resources of the Nile River, in: Climatic Change 46, Nordrecht, pp. 339–356.

Welthungerhilfe (2007): Herausforderung Hunger: Welthungerindex, Fakten, Ursachen, Entwicklungen, Bonn.

Whittington, D. and McClelland, E. (1992): Opportunities for Regional and International Cooperation in the Nile Basin, in: Water International, 17:3, London, pp.144-154.

World Bank (2000): Can Africa Claim the 21st Century?, Washington.

INTERNET

World Bank (2010): World dataBank,
URL: http://databank.worldbank.org/ddp/home.do#ranking, retrieved: 10-04-10.

Egypt State Information Service:
URL:http://www2.sis.gov.eg/En/Economy/Sectors/Industry/050303000000000001.htm, retrieved: 15-04-2010

Research and Markets: Egypt Tourism Secotr Analysis
URL:http://www.researchandmarkets.com/reports/543809/egypt_tourism_sector_analysis, retrieved: 19-04-2010.

CIA World Fact Book:
URL: https://www.cia.gov/library/publications/the-world-factbook/geos/tz.html, retrieved: 16-04-2010..